The Contractor's Closing Success Blueprint

A Contractor's Guide: Consistently Close More Profitable Jobs and Generate Higher Quality Leads

By
Mike Jeffries

© 2014

Mike@ClosingSuccessSystem.com

Published By
MAP Business Growth Strategies, LLC

The Two Biggest Problems Contractors and Service Companies Have When It Comes to Generating the Most Profit for Their Business

(Which one is Bugging the Heck out of You?)

The #1 Problem most companies have is this: They are Closing *Less* Than 50% of Their Sales Calls.

Symptoms Include:

- Tough to Schedule Appointments When the Prospect Contacts You
- Prospects Only Seem Interested In Price
- Cancelled Appointments at the Last Minute
- Lowering Your Price Just to Get the Job
- Losing Jobs to Competitors that You Know Aren't as Good as You Are

The #2 Problem I hear all the time is this: You're Spending Good Money on Advertising but Don't Seem to be Getting Much Bang for Your Investment.

Symptoms Include:

- Poor Quality Leads
- Price Shoppers
- Your Website and Ads Pretty Much Say Exactly the Same Things as Your Competitors
- Your Only Call to Action is "Free Estimates"

My name is Mike Jeffries and I am the managing partner of Rivers of Revenue, LLC. I've worked exclusively with residential and commercial contractors and service companies since 2002.

I've created a simple to implement, easy to understand, complete program that will allow you and your sales team to, **first** - increase your closing rate by 10%-50% almost immediately and **second** - generate <u>more leads and better quality leads,</u> from your current website, ads and marketing, without having to spend any more money than you're currently spending (or planning to spend).

Preface

I wrote this book because the primary goal of my company is to help thousands of contractors and service providers across the world to close more sales.

It is my sincere belief that this book will help many people in the contracting and service industries to do just that.

I created the original version of this book as a training and motivational tool for my one-on-one consulting clients.

It helped many of them understand how to communicate the value of their company in very specific ways that buyers instantly get.

Since 2002, I have been blessed to work with many fine contractors and service providers who operate their businesses professionally. They have taught me many things and have helped me to become a better consultant and a better listener.

Over the years I have had the privilege to share what I know and have learned at industry and supplier conferences on the national, regional and local level. I am an Instructor for the Painting and Decorating Contractors of America (PDCA) and have been honored to be a member of the Panel of Experts of the National Association of the Remodeling Industry (NARI).

The rapidly changing nature of technology has allowed me to educate contractors through webinars, my Quick Tips© blog and my Contractor's Marketing & Closing Success Blueprint Podcast©.

Contractors across the United States, Canada, Australia, South Africa and many more places have become followers and fans simply because they were able to find me through search engines, discussion groups, iTunes, Google+, YouTube, LinkedIn and other social media platforms.

The Closing Success System© and the Contractor's Closing Success Formula© have been successfully implemented for contractors and service providers in all of these industries:

> painting - home remodeling - commercial and residential renovation - roofing - siding - windows - landscaping - decks - marine construction - fencing - custom closets - flooring - millwork - plumbing - power washing - log home restoration - home theater - electrical - property management - cleaning - information technology and trucking.

You may find yourself thinking as you read and review certain sections and chapters that this is just too basic to be effective. Let me assure you that even the very best companies we have had the privilege of working with had only implemented a handful of these strategies and tactics consistently before they started working with us.

So take the time to implement and I am confident that you will be amazed at the number of buyers that will now say yes.

Please tell me about your successes and if you are struggling let me know that too. Also, please share your ideas to make this better. You can email me at Mike@ClosingSuccessSystem.com.

I wish you great success in your business and in your life.

Mike Jeffries – June 2014

PS: I am very interested to know what you think. After you finish this book, I would appreciate your feedback with an objective review on Amazon.

Table of Contents

Foreword   10

Introduction   13

Introduction to Part 1
Your Marketing Foundation©   18

Section 1
Are You Closing Less Than 50%?   19

Section 2
Seven Major Trends That Are Impacting Your Sales and Profits   22

Section 3
Why This Program Increases Your Closing Success and Lead Generation   31

Section 4
Where Most Companies Drop the Ball...And Lose Profitable Jobs. Are You Guilty of This?   34

Section 5
The Only Real Way to Close More Jobs...Without Lowering Your Price   36

Section 6
The Secret That Will Allow You to Close More Business...Almost Effortlessly   38

Section 7
If You Ignore This Essential Step Your Prospects Would be Crazy *Not* to Insist on a Lower Price!   40

Summary Part 1
Your Marketing Foundation©   42

Introduction to Part 2
Your Sales Closing Success
Implementation Plan©   43

Chapter 1
The One Characteristic All Successful
Contractors Have   45

Chapter 2
Getting Your Employees and Crews to
Supercharge Your Closing Success   49

Chapter 3
How to Win Over the Price-Focused Prospect   55

Chapter 4
Setting Yourself Up For Closing Success   60

Chapter 5
Getting Prospects Ready to Buy   68

Chapter 6
Delivering the Number   73

Chapter 7
Four Words You'll Begin to Hear More Often:
"When Can You Start?"   78

Chapter 8
Successfully Overcoming the Price Objection
and Other Common Objections   86

Chapter 9
Know When To Follow-Up and
When to Walk Away   97

Summary Part 2
Your Sales Closing Success
Implementation Plan©   105

Bonus Chapter
Marketing's New Rules™ of Closing Success   106

The Next Steps   117

Resources   123

Foreword

Passion for one's beliefs and experiences is a cornerstone for the motivation of success. Over the years I have co-founded various software related businesses focusing on both small and large businesses. And along the way I have met extraordinary people who helped me shape my own knowledge within each industry. My latest venture targets the small business contractor in the construction industry who wants to learn how to get more leads and manage their jobs more efficiently. One of the people who I am happy to have met in our industry is Mike Jeffries of the Closing Success System. Mike offers not only successful guidance to help contractors, but he possesses the desired passion to offer the best he can to help contractors evolve as better businesspeople and not just craftsmen.

I met Mike a few years back after another well-established consulting group mentioned his name more than once. In addition to the reputation of that consulting group, I have had numerous prospects and customers mention they either worked directly with Mike, or subscribed to his Do-It-Yourself education program focusing on better sales closing success. It is absolutely great when you hear nothing but different levels of praise and success – without any negativity.

My own personal experiences with Mike have been working with the same clients, but with synergistic services. There has been a little service overlap in our desires to help contractors get more work and manage the job – but the vast majority of experiences have been consistent in quality, professionalism and cost effectiveness. I have also had the pleasure of helping Mike with his own online marketing initiatives, and while Mike may already have known about many of the topics I introduce, he has always been receptive to making good things better. It shows a lot in ones' character when they are willing to learn more, even when they already know a lot.

Whether it be Mike's writings, podcasts, blog updates or Do-It-Yourself Educational Series, the biggest shortfall I can see is when someone passively looks to learn but doesn't take the initiative to make actual changes. You will not lose any money just passively reading Mike's book but you will lose a lot of opportunities by not taking the advice and implementing the strategies and tactics presented.

Regarding this book, you will find it enjoyable to read, straightforward – and most important – easy to retain ideas and implement them. Even if you get just a few suggestions mastered and make a few more part of your daily business arsenal, I am sure you will obtain more business over a period of time. Most contractors are a craftsman first and then a businessman. The concepts of this book will help you in your journey to help you marry the importance of your business with your craft.

I encourage all in the industry to have this book be one of the handful of books they keep on their reference shelf.

Wishing you great success with the growth of your business.

Brian Javeline
President & Co-founder
Servusxchange, LLC., creator of the MyOnlineToolbox.com contracting platform

Introduction

I'd like to welcome you to this program. In it I will teach you what you need to know to consistently close profitable business and how to generate higher quality leads.

I am an instructor for the Contractor College of the Painting and Decorating Contractors of America and I have been honored to be included on the Panel of Experts for the National Association of the Remodeling Industry. I have presented our marketing workshops and webinars for professional associations and suppliers in many contracting and service industries on the national, regional and local level.

But this isn't about me, this is about you and how I can help you grow your business, close more jobs and make more money. No single strategy or tactic in this program will make you successful but *taken together* all these little (and big) things will practically guarantee that you will be more successful.

You may be thinking as you read: "Come on that is too simple to be effective or sure I have heard that before but I can't believe anyone actually does it." I can assure you our clients do these tactics and strategies and they are motivated by their initial success to do them consistently.

It's basic blocking and tackling. In fact, as you review and study this program you may find yourself saying,

"Hey...we already do that". Chances are you probably are doing some of the things in this program. But here's the difference. You either don't do them all...or you don't do them consistently...or in the right order...or with the wording that we've developed through testing and implementation since 2002.

You may also think, "Gosh Mike that is common sense". Common sense is not common practice. I thank you for making the commitment to invest in this program, now it is time to make the commitment to put it into practice.

This program has been in the making for well over ten years. I am going to share with you the very best practices of successful companies just like yours. These are the same practices that we teach all of our clients that use the Closing Success System© and the Contractor's Closing Success Formula©.

This program has two parts.

1. The first part: "Your Marketing Foundation©" has seven sections.
2. The second part: "Your Sales Closing Success Implementation Plan©" has nine chapters plus a bonus chapter.

We recommend that you review the entire program and then go back to those sections and chapters that you want to review again. The more you study AND implement the better you will become.

How many times should you review it? How bad do you want to win?

I have been told by a very knowledgeable person, that a *professional football player spends 500 hours in practice, study and training for every hour on the field in a game.* Wow 500 hours to 1.

Now I am not suggesting that you need to review this 500 times. What I am suggesting is that you read and re-read the entire program at least ten times. Then create the tools and practice the scripting with prospects until you know this so well that you can deliver it naturally in your own style.

In a nutshell, what I plan on sharing with you are proven, field-tested tactics that we've developed for our clients that will allow you to:

- Take all the guesswork out of your presentations and ensure that you can confidently overcome the most common objections
- Know what to say before, during and after meeting with the prospect
- Give you the confidence to close any profitable opportunity
- Simply and easily increase your closing success without learning any slick sales training techniques or closing tricks
- Increase your quality referrals and repeat business opportunities

I don't know of anything in life that is successful that isn't based on a system. Your competitors fail big time because they wing it and don't have a system to follow. Our clients succeed because they consistently follow a system and now you will too.

I want to emphasize that this program isn't just about what you say when you meet the client. It covers everything you say, do and show the client from the time they initially contact you - all the way until they give you a check and a referral to a friend – and everything that MUST happen in between if you want to get their business.

Why should you spend time getting good at this?

If you are like most of the companies we work with, you are closing 20% of your proposals or one out of every five.

If you just move your closing rate up 5% from 20% to 25% - that would jump your business 25%. If you are doing around $400,000 a year that would be a $100,000 increase in sales every year just for closing one more prospect out of twenty!

Your profit on the additional $100,000 will be significant – most likely at least 25% - unless you need to add capacity (and most companies don't to handle a 25% increase) then all you'll need to pay for is material and labor.

Then you will only need to pay for material and labor. So what would you do with the extra profit every year? How about getting out of debt or taking a dream vacation or saving for college or retirement...put your goal on your wall where you will see it every day!

Now you should be really motivated to get started.

One more important note – failure to implement is the #1 reason that marketing programs fall short of expectations. Make a commitment to put these strategies and tactics in place and start today.

Introduction to Part 1

Your Marketing Foundation©

In Part 1 of the Contractor's Closing Success Blueprint© I will give you knowledge and understanding to build Your Marketing Foundation© so your company can prosper in any economy.

You will learn:

- Why you are leaving a ton of money on the table
- The seven major trends that are affecting your business so you can take advantage of them and increase your profits
- How to think like your buyer
- Why you are losing some jobs to unqualified contractors
- Why you won't need to compete on price
- What prospects really want to know about your company and your services
- How to make your company the obvious one to do business with

Once you complete this part you will understand more about closing and lead generation than any of your competitors.

Section 1

Are You Closing Less Than 50%?

What would it mean for your business if you could book all the business you could possibly handle?

Instead of closing two out of ten calls, you – or your sales team – would come back with contracts on almost EVERY prospect call they went on.

No longer will the profitability of your business rest on the shoulders of one GOOD sales guy (maybe that's you).

You have at your fingertips the tools to make an average salesperson GOOD...a good salesperson GREAT...and a great salesperson a SUPERSTAR.

And to do this, you won't have to spend thousands of dollars on "sales training" or consultants who know nothing about your industry or a new advertising program. In fact, you wouldn't have to make any major changes to the way you do business.

All you'd have to do is implement an often overlooked and critical step in the buying process that's based on common sense.

With our easy to implement system - when your sales rep shows up for the sales call, your prospect already believes that your company is the obvious choice to do business with - *your competition never really has a chance.*

Now, instead of rolling the dice and throwing countless dollars into advertising, trying to generate more and more leads, riding your sales guys to "do better" and worrying about where the next project will come from, you'll have more than enough business.

You and your sales reps will be excited about going out on prospect calls, because you will all know it's probably going to put money in your pocket and your sales rep's pocket.

You will no longer need to lower your price (i.e. your profit) to get a job because in the prospect's eyes, they realize that they would have to be nuts to choose another company…regardless of price.

Your installation crews are busy and getting compliments from your clients.

And now your biggest problem is deciding how much of the extra profits to sink back into the company and how much to keep for yourself.

I've worked with residential and commercial contractors and service companies – from an owner and one helper to companies with 200 employees.

And here are three VERY important facts I know about contracting and service businesses and more importantly, your prospects:

1. The average closing ratio on "non-referral" leads is almost always less than 25% and often as low as 8%

2. Most even struggle to close 50% to 60% of referrals
3. 70% of the clients that don't immediately close *will buy within the year... from someone else*

Here's what these facts tell me: *You're Leaving a Pile of Money on the Table* simply because you're not closing as many of the prospects as you deserve.

Section 2

Seven Major Trends That Are Impacting Your Sales and Profits

I recommend that you spend some time reading and thinking about the information in this section. This will put you far ahead of your competitors.

You have probably noticed that your marketing just isn't working the way it used to and you aren't sure why.

The explosion of the internet and the dramatic increase in the power of mobile devices has changed the marketing game for contractors forever in just the last five years.

The good news for people reading this book is that most contractors haven't figured this out and they are still doing things the same old way even if it isn't working like it used to.

Trends You Need to Understand and Use to Your Advantage

1. There aren't any surefire lead generation strategies.
2. Prospects can easily find alternative providers of your services.
3. Prospects have an unprecedented ability to check out your company before contacting you.
4. Buyers now control the sales process.

5. Prospects and clients are comparing their experience with your company with their experience with other outstanding local and national companies.
6. Personal relationships are vital to your success in an age of social media.
7. Quality websites are an important first step in closing.

All of these major trends have impacted your contracting or service business even if you have continued to grow in the last five years.

Once you understand these trends and how you can use them to your advantage you will move light years ahead of your competitors.

I will review each of them in order.

1. There Aren't Any Surefire Lead Generation Strategies

Think back just a few years – as a local contractor or service company you could place an ad in a directory like the Yellow Pages or the Blue Book and it would generate calls.

Direct mail and advertising in local print publications were also tactics that worked. One remodeling contractor I know would send out 40,000 postcards in the spring and fill up his whole year just based on that one mailing.

Most contractors have eliminated directories from their marketing strategy or they only participate in web-based versions. The results from web-based directories pale in comparison to the "good old days".

All of these dying methods are examples of outbound marketing – advertising to all buyers in a niche or in an entire market.

Today direct mail is rarely used and if it is used the results are typically breakeven at best. Watch your mail for the next month and see if you see any stand-alone direct mail pieces from contractors or service providers.

Newspapers are folding or laying off in an effort to survive. Niche print publications rule the day and broad-based magazines have moved to an online platform.

Prospects now get their news and information streamed to them instantly or they can do a quick search for answers.

Inbound marketing is the most successful approach now.

This means that you have to set up your marketing so buyers can find you when they have a need.

You must have multiple smaller marketing tactics in place because there aren't any homerun solutions that will fill up your calendar.

Now you may be thinking: "Aren't internet searches the same thing buyers did when they looked in the Yellow Pages?" The answer is yes to a very small extent. Most ads were simply a listing of services or products offered. Directories offered a very limited way for people to learn about your company.

Today buyers have instant access to just about any information they want about the product or service you offer and most want to do their own research before contacting a supplier of a service or product.

You need to be ready.

2. Prospects Can Easily Find Alternative Providers of Your Services

In the past a prospect had to meet with the company to learn what they did and what services they offered. The meeting was also the place where the company would sell the prospect.

The prospect had to endure the sales presentation and often felt pressured to buy even if they didn't like the company.

Now prospects can easily find alternative suppliers with just a few clicks even if they are searching for a niche service. This ability to quickly search makes it imperative that companies be in the game with a strong online presence.

Now prospects don't have to meet with you to learn about your company and I will cover that in the next point.

3. Prospects Have an Unprecedented Ability to Check Out Your Company Before Contacting You

This is the biggest change in the landscape of marketing.

Prospects can review: your website, social media sites, rating companies and a myriad of other sources to find out about your company and what you offer.

They no longer have to meet with you to learn this vital information.

They can review your competitors the same way.

Plus they can find out what to look for, what to look out for and how to hire a qualified contractor or service company.

All of this information is available without meeting anyone.

With all of this information and education available it becomes a much easier task to eliminate companies that have a weak online presence.

As a base point a website that clearly sets the standard will put you miles ahead of your competitors. Protecting your image on ratings sites is becoming increasingly important and there is legal action that seeks to change what can and can't be posted without "proof".

Without a strong online presence you may never get to bat even if you were referred to the prospect.

All of this access is the primary reason the buyer now controls the sales process.

4. Buyers Now Control the Sales Process

Buyers can now learn about you and your competitors. They can also learn what to expect when they buy, what constitutes a good deal, what to look out for and so much more.

They now control the information about the product or service they want because of all the information they have access to.

They also control the decision to meet you or not meet you and they control the final decision on whether to buy or not.

All you can control is the information about your product or service and how it compares to your competitors and industry standards.

Your advantages better be very clear and focus on what the buyer wants to buy NOT just what you have to sell. Create your information to answer the problems and frustrations your buyer is looking to solve.

If the information about your company looks and sounds like every other competitor you will be left to compete on price.

5. Prospects and Clients Are Comparing Their Experience with Your Company with Their Experience with Other Outstanding Local and National Companies

People now share their experiences with the world on social media.

It is very easy to learn about someone's great or not so great experience.

You can find out what hundreds or thousands of people have experienced and think on social media sites. I definitely check out Trip Advisor before trying a new restaurant because I can. I have also found that by reading the good and not so good ratings you can get a good idea of the consistency of service.

Based on all of the information available, your expectations will be set or at least influenced.

I also want you to think about companies that have changed their industry – Disney, FedEx and Dell to name a few. They set the bar that every other competitor has to match.

Let me give you a quick example – Dell set the bar – you could contact them and get a fully customized computer delivered to you the next day or in two days.

Now if you were looking to order custom windows for your home, you may be wondering why it takes weeks to get them in comparison to getting a new computer.

So Dell can influence what window company you might choose – one that is able to deliver custom windows like Dell delivers custom computers is very likely to win your business.

If you buy coffee at the same place every day, they most likely know what you want before you order it.

We patronize local companies that meet or exceed our expectations.

Think about the companies you buy from and why you do it and it should be pretty clear if you need to make some changes to up your game.

6. Personal Relationships Are Vital to Your Success in An Age of Social Media

I am sure you have heard of people that have hundreds of friends on Facebook. It is very easy to have a virtual relationship with someone online. It is easy to connect without committing to a real relationship – like talking at a party. Most of these relationships are superficial.

I am also sure that if you are looking for a recommendation, that you will ask close friends or business associates rather than putting it out to all of your online friends.

The message I want to deliver is that personal relationships are vital today for referrals and endorsements. These relationships have become the backbone of lead generation. Strong joint ventures or cross promotion relationships will help you achieve a steady flow of quality leads.

7. Quality Websites Are An Important First Step In Closing

All of the key trends I have already mentioned should have made it clear that a strong website is a basic building block for lead generation and closing.

This is the place that you can control the information and education you want the buyer to have.

The great thing is that most contractors either don't have a website or it is so generic that it is just a tri-fold brochure online.

You can definitely gain a significant advantage by upping your game.

Now that you know about these major trends it will make it much easier for you to take advantage of the opportunity to make your company stand out.

Action Step – Read the rest of the book and then come back to this section and read it again. When you do that it should be clear how the action steps will help you close more business and grow your profits.

Section 3

Why This Program Increases Your Closing Success and Lead Generation

There's an age old saying in the sales world:

If you want to know why John Smith buys what John Smith buys, you have to see the world through John Smith's eyes.

This was true eighty years ago and it's just as true in today's high-tech marketing world.

What does that mean to you?

First you have to understand a little bit about how "John Smith" thinks when it comes to companies like yours:

(Please don't take this personally...)

- Like it or not, the perception of your industry is that you're just a bunch of "bubbas" or "two guys and a truck".
- You and your competition pretty much sell the same services and there's no real difference from one company to another.
- There is the impression that because all the websites, ads and marketing materials say the same stuff: family owned business, in business since 1972, high quality materials, licensed and insured etc., etc., etc., *that you are all the same!*

- Your prospects hear this same general pitch from every owner or salesperson: "we give great service", "we're owner-operated so we care", "we train our people thoroughly" and on and on.

Is it any wonder that almost all prospects draw the conclusion that all contracting companies are the same – so price must be the best way to make the choice?

I'm not saying any of this is true about your company. But it *is* the *perception* that the prospect (John Smith) has about your industry.

Don't believe me? What do you think of when I say "used car salesman"? You do have a perception – true or not?

Unfortunately, your prospect John Smith has the same perception when they are looking for a contracting or service company.

Here is the best thing about this program – you don't need to make wholesale changes to your ads and give your competitors a clue about what you are up to.

First, the Closing Success System© and the Contractor's Closing Success Formula© focus your efforts on closing prospects who are already calling or contacting you.

Increasing your closing rate is by far the easiest and fastest way to build your backlog and make more money.

Prospects that are calling you are ready to buy.

Now you will have the tools to make your company the obvious choice.

Once you do that, then you can invest some of your new profits into improving your lead generation if you need to.

Many of our clients have doubled and some have tripled their closing success.

At that point they realized they didn't need more leads so they concentrated on improving the quality of all of their leads.

Section 4

Where Most Companies Drop the Ball...And Lose Profitable Jobs. Are You Guilty of This?

Here's an all too familiar scenario...

You get a phone call from a prospective client named John Smith who's interested in getting his kitchen remodeled (or his house painted or a new roof installed or landscaping updated)...

After a short conversation with John to get a feel for what his needs are, he asks the dreaded question: *"So how much do you think it will cost to do the job?"*

You do your best to tell John that you'll have to evaluate his specific job and cost it out and then give him a quote.

John seems to understand, and says he'll wait for your bid. You set up a time to go out to John's place to take a look.

Meanwhile, John calls around to four or five other contractors and goes through the same process.

If you're lucky, John remembers he set an appointment and he's home when you show up. You show John some ideas of what's available. Ultimately, you submit your quote to John and begin the wait.

As you wait, you wonder if your quote is low enough to land the job AND at the same time you wonder if your quote is high enough to make any money.

You wonder why it's been three days since you submitted the quote and you haven't heard back yet.

Finally you call John only to hear him squirm as he tries to break the bad news gently: *"Oh I'm sorry, we've decided to go with another contractor."*

You try to ask John why, but since he's polite - and since he's extremely uncomfortable talking to you - he tries his hardest not to give you a direct answer. He probably says something about another contractor and lower prices or a friend of the family...

Let's face it: there are really only two possible reasons why you lost the job, assuming you followed up in a timely fashion: *either your price was too high or John Smith's confidence in your abilities was too low (he couldn't figure out what was different about what you offered – in other words the value).*

In the worst situation it could be both.

In any event, you now have to wait for the phone to ring so you can live through the whole frustrating scenario all over again.

Section 5

The Only Real Way to Close More Jobs...Without Lowering Your Price

Is there any way to avoid this situation?

Is there a way to let prospects know exactly what you do that makes you better than your competitors?

Can you get away from always having to compete on price?

And come to think of it, is there a way to get that phone to ring more often? And when it does ring, wouldn't it be nice if you could somehow get your prospects to say to themselves: *"I would have to be nuts to choose another company...regardless of price!"*

Here's the good news: in a world of intense competition, rapidly rising advertising costs, increasingly price-sensitive consumers, and seemingly less time to sort through it all...there is a way to *TAKE CONTROL* of your business and *start landing more jobs at higher margins and that means you make more money without doing more work.*

What a novel concept and you may not be ready to believe it is possible just yet...

Let me point out that every contractor I have worked with swore up and down that price was the only thing that mattered to prospects.

That's what they thought when we started to work together. Now if you ask any of them – price means nothing more than 80% of the time. Of course there are still that 20% that will make a price-based decision despite all of the reasons not to do that – let them - you wouldn't be making money on those jobs anyway.

If you are still convinced that price is the most important thing to every prospect you can toss this now. Please don't blame me if a direct competitor of yours calls me and implements the program.

Do you think that this sounds like a pipe dream?

Actually, you could be living your dream if you just learn a few simple principles.

Section 6

The Secret That Will Allow You to Close More Business...Almost Effortlessly

If you want to start closing more of the calls you go on, you must memorize the most Important FACT about your average prospect:

> *95% of the prospects who are calling you have never remodeled their kitchen (or installed an irrigation system or put a roof on their home or painted the outside before)* or if they have it was a long time ago and they weren't happy with the project – otherwise, why wouldn't they be calling the original contractor back?

You see there are really only two reasons why you are getting a call from a new prospect:

> #1 - They hired someone in the past and weren't happy
>
> #2 - They don't know anyone or anyone they are ready to hire

And because it's such a rare purchase your job is NOT to sell them on your company.

Let me repeat that...

Your job is NOT to sell them on your company!

Your job, initially, is to *TEACH* your prospect *how* to get their project completed professionally, without getting taken advantage of.

- What to look for
- What to look out for
- What they need and don't need and why

Once you do that, then you will show them what they should expect from a *quality* company (like yours) when it comes to: service, materials, workmanship, guarantees and follow-up.

This is what I refer to as "letting them buy" and you will learn a lot more about this as you read on.

Section 7

If You Ignore This Essential Step Your Prospects Would be Crazy *Not* to Insist on a Lower Price!

When you start educating your prospects, up front, while they're actively seeking a contractor or service company, you'll be positioning yourself and your company as the industry expert that all other companies must be compared to.

In the eyes of the prospect, you'll become the Standard for your industry – and the Obvious Choice to Do Business With.

Knowing this...and by taking full advantage of this opportunity, allows you to position yourself as the obvious choice to hire.

Simply Stated: You'll be closing more of the prospects you meet, easier than you ever imagined and rarely – if at all - will you have to lower your price to get the job.

In fact, companies that have implemented the Rivers of Revenue Closing Success System© or the Contractor's Closing Success Formula© have actually *RAISED* their prices without a measurable drop off in closing success.

Hmmmm...So now you're closing more jobs for a higher price.

Are you starting to see how this program can positively impact the profits in your business?

Even if you don't think you can raise your prices now...you will once you start to close more jobs.

So how can you do this in your business...starting today?

Now that you have a solid marketing foundation in place I can now give you Your Sales Closing Success Implementation Plan©.

Summary Part 1

Your Marketing Foundation©

The explosion of information and education available on the internet has given buyers control of the sales process.

The good news is that virtually all contractors are still doing business the same way they always have.

Implementing straightforward changes in your marketing will give you a huge advantage over your competitors and make it easy for your prospects to choose your company.

The biggest reason that marketing programs fall short of expectations is because they never get implemented.

In Part 2, I will teach you how to implement your sales closing success plan so you can close more jobs and make more money.

Introduction to Part 2

Your Sales Closing Success Implementation Plan©

In the Introduction to this book, I explained that failure to implement is the #1 reason that marketing programs fall short of expectations.

In Part 2 of the Contractor's Closing Success Blueprint© I give you an easy to execute implementation plan so your marketing program will exceed your expectations.

I'll take you out into the field. I will walk you through the initial prospect contact, your prospect meetings, delivering the number confidently and overcoming the most common objections especially the price objection.

Every tactic and strategy comes right from our client files.

You will learn:

- The most important characteristic of the successful people we have trained
- Why most marketing programs fail to meet expectations and how to fix that
- How to overcome your prospect's biggest frustration
- What to say…and NOT say…when you meet the prospect
- Delivering the number confidently

- How to overcome the price objection
- What you need to show your prospect to become the obvious choice
- Proper follow up - without being a pest - and knowing when it's okay to walk away

There are nine chapters in Your Sales Closing Success Implementation Plan©. At the end of each chapter I give you *specific* action steps to take to implement the plan into your business.

In a special bonus chapter I teach you what to do to dramatically increase your success. Most contractors are still doing things like it was 1985 and you will learn how to take advantage of them so your business will prosper now and for many years to come.

Chapter 1

The One Characteristic All Successful Contractors Have

If I could give you the one characteristic of the successful people we have trained it's this – they believe in themselves and the value they offer.

They believe that they offer the best value in the market and they know that they always strive to give high value to every client. They simply can't understand why every prospect wouldn't want to hire them.

If you don't believe in yourself, no one else will. Belief in self isn't being arrogant. This is simply the confidence that what you are offering is worth every penny and that the client will be very satisfied in the end.

Our clients are also confident that they will be enthusiastically referred by every client.

I have always played a lot of sports and right now golf is my passion. I can assure you that if I am standing over a shot and thinking: "I hope I don't hit this into the water" that is not going to help my confidence. With that thought in my head – it is much more likely that I am going to fail than succeed.

I still hit some poor shots, heck every professional does. But now when I step up to the ball, I know what I need to do, I envision where I want to hit it.

The pros call that visualization or picturing what you want to happen. It simply helps your brain direct your body to do what you want it to do.

The pros have also done something else – they have practiced for many, many hours to get the confidence that they can execute each shot. Interestingly, you can never practice every possible golf shot you will face because there are an almost infinite number of places your ball can end up on the course.

But you can practice a lot of similar shots so you'll have the confidence that you can do it. This is the ability to adapt to a new but similar situation. *I refer to that ability as lateral thinking.*

That is another key skill of our successful clients, they have practiced and they don't panic when a prospect asks them a question they haven't faced before. Preparation breeds confidence and belief in oneself. (I will discuss lateral thinking more in Chapter 8.)

How Can You Build Belief in Yourself, Your Product and Your Service?

Here are seven action steps to implement:

1. Make it your personal goal to get a great testimonial from every client. The more you hear what a great job you do, the more you will believe it.
2. Require every client to complete a satisfaction survey so you can find out where you need to get better and then put a plan in place to fix whatever needs to be fixed.

3. Learn the best practices by taking additional training in your craft from suppliers to your industry and professional associations for your industry.
4. Take your best clients to lunch and ask them: "Is there anything we could do to improve our service or was there anything we could have done to improve our service?"
5. Ask your employees what they would do to improve your product or service if it was their company.
6. Practice saying every morning as you head out the door: "I am the best damned contractor in my market and today I am going to prove it." Naturally– fill in your profession.
7. Practice what you are learning on this program until it becomes automatic and completely natural.

I have one more key recommendation for you: read a self-help book.

There are thousands of self-help and motivational books on the market. I would start with any book by Dale Carnegie or Napoleon Hill – virtually every self-help and motivational book on the market today is based on what they wrote 50 or more years ago, *A Year of Growing Rich* by Napoleon Hill started me on my path to success.

If you don't believe in yourself, no one else will. Start believing in yourself today.

Visualize the prospect saying every time – "When can you start?"

Action Steps – Here are three action steps to put into place:

#1 – Make a plan to complete each of the seven action steps already mentioned. Most important, get your testimonials together and create a client satisfaction survey and get it into place.

#2 - If you send me an email at Mike@ClosingSuccessSystem.com I will send you a proven client satisfaction survey that is ready for you to use. This is the same one our clients have used for years.

#3 - Finally, get a motivational book by Dale Carnegie or Napoleon Hill and make it a daily exercise to read for 15 minutes and then implement one idea a day.

Chapter 2

Getting Your Employees and Crews to Supercharge Your Closing Success

I hope you took the time to read the Foreword, Introduction and Preface because each section includes great information and advice you won't find anywhere else in the program.

In the introduction I included a very important note – the failure to implement is the #1 reason that marketing programs fall short of expectations.

The #2 reason that marketing programs fall short of expectations is because employees and crew members don't embrace the program the owner or supervisor wants to implement.

This is really a management and ownership failure and you can fix it by standing tall in your belief in the program. If the ownership takes a "do as I say but not as I do" approach the resulting failure is very predictable.

There is a fundamental human attitude that you need to control and that is the fear of change. A large majority of the population fears change and you can bet your employees or crew members have this fear.

People get comfortable and they know their routine and they may even resent the change.

As the owner you need to set the tone that implementation is going to make life for everyone better even if it is a bit painful at the start.

How many people do you know that have a hundred reasons why something won't work and not one suggestion on how to make it work?

You have to think to yourself, "How much is this costing me?"

A Short Case Study in Lost Profits

Let me give you an example from my client files.

I am sharing this not as self-promotion. I want you to be able to analyze your business and to believe that you can make a lot more money if you have the courage to change.

This contractor had six sales reps on his staff and they were all pretty young and the four that had been with him the longest were making great money. In fact they were probably making two or three times the money of most of their friends. I will refer to them as the "comfortable reps".

The two newest reps were doing pretty well but didn't have enough of a track record to produce credible numbers. I will refer to them as the "newest reps".

On the surface it looked like the team as a whole was doing well because they were closing about 45% of all their leads. When we ran the numbers we realized that 50% of their leads were coming by referral and they were successful about 70% of the time on those leads.

The other 50% of their leads were "cold" leads generated in a variety of ways and they were only closing 20% of those leads. So overall it looked like they were doing great but when a prospect wasn't referred to them they were right there with everyone else – losing four out of five proposals to competitors.

We implemented the Closing Success System© and the two newest sales reps bought in fully. After a couple of months the newest reps were closing 15 percentage points higher than the comfortable reps. Not 15% but 15 percentage points higher – overall they were closing 60% of their leads. I will tell you that this type of success is a bit unusual but they are not our only clients to reach these levels.

Even though I had worked with all the reps, the comfortable ones just weren't buying in and the owner wasn't pushing them.

I did a little math for him and that made him see the light.

As a team, each sales rep was generating proposals in the $2,000,000 range for the year. Their average overall closing rate of 45% rate meant that they were generating approximately $900,000 in sales.

However, the newest reps were generating $300,000 more business on the same proposal base.

This meant that the four comfortable reps – as a group - were letting $1,200,000 in new business slip through their fingers.

You can do the math to figure out how much profit the comfortable sales reps were costing the owner.

I recommended that the owner let two of the comfortable reps go and replace them with people that would buy in fully to the system. He was very reluctant to do this since they had been with him as his business had grown and I am sure he liked them.

In the end he did let two of the comfortable reps go but it cost him a ton in profit since he waited almost six months to do it. He "hoped" they would be inspired and change but they didn't.

This owner kept good statistics on their closing success so it was pretty easy to see who was doing well and who wasn't.

Unfortunately, most of the contractors and service companies we have worked with haven't done such a good job on this.

When I ask a prospective client what their closing success is most say something like, "Well I think it is around 20%".

If you don't know what your closing rate is on each type of lead you won't know what is working and where the best opportunities are.

The Often Overlooked Statistic – Volume Closing Rate©

There is another statistic you should be tracking and that is your "volume closing rate©". This is something even fewer people track but they should. The volume closing rate© is the percentage of the total proposal dollars that you close.

Why is this important?

Many contractors have good success on smaller proposals but are below average on the larger projects. One of our most successful clients was closing almost 70% of all proposals when they came to us for help. I suggested that they really didn't need my help if they were landing seven out of ten proposals.

The owners told me that even though they had an impressive success rate overall – on large proposals over $100,000 they were only landing 33%. They couldn't understand why.

Once they implemented the Closing Success System© their success rate soared to over 50% on the larger proposals and with it their volume closing rate© got very close to their overall closing rate of seven out of ten.

Action Steps – Here are four action steps to put into place:

1 – Determine your overall closing rate and your volume closing rate© for each type of lead.

2 – Calculate the impact on your profits if you increased your closing rate and your volume closing rate© by ten percentage points. For example: if your overall closing rate is 24% calculate what your sales and profits would be if you can increase it to 34%.

#3 – Use the statistics to educate your entire team on their success by posting them on the wall of your office.

#4 – Captain your ship to success by putting the statistics in your office where they will be a constant reminder of the impact on your business of implementing the needed changes in your business.

Chapter 3

How to Win Over the Price-Focused Prospect

If you want to be more effective at closing sales, you need to understand what the buyer wants so you can give it to them.

My company has done countless surveys of your buyers to find out what they really want. Professional associations and suppliers to your industry have also done the same type of surveys.

Let me share the most important responses. Some of the answers may surprise you. Let me assure you that the answers we receive have not changed since I started helping contractors in 2002.

Here are the eight biggest frustrations buyers tell us about contractors and other companies in service businesses:

1. They never call back
2. They never show up for appointments on time
3. They don't start or finish the job on time
4. It always costs more in the end than in the proposal
5. It won't look as good as they claim or work as well as they promise
6. They don't honor the warranty or guarantee
7. They don't finish the punch list

Here is the most important frustration and you need to understand this about your buyer:

> 8. They desperately want to hire a trustworthy company but they don't know how to do it

If I can summarize these frustrations – buyers are talking about communication, respect and reliability.

I want you to realize that price will always have some impact on the sales process BUT if you can clearly demonstrate through your actions that you communicate consistently, respect them and their time and that you are reliable, price will essentially be eliminated from the decision over 80% of the time.

You may still believe that price is the only thing that matters. If that were the case then wouldn't the low bidder win every time? Wouldn't we all be driving $10,000 cars? Wouldn't Starbucks be out of business?

One more point – it is shocking to me that contactors don't call back or that they call back only when it is convenient for them.

Let Me Explain Why Price Seems So Important to So Many Prospects

When all the marketing and advertising looks and sounds the same to the prospect, they believe that they can hire anyone and get the same result.

When that happens price is the logical way to choose.

In our experience virtually all the marketing and advertising does look and sound the same and that is why most people believe that price is the most important factor.

Now you know why you need to look and sound different from your competitors to avoid the price battle.

Let me give you the key. Be specific.

Let me repeat - be specific in everything you claim about your company, service and product in all marketing and advertising.

We will talk about this more in later chapters but here are a couple of examples.

"They never call back" is a huge frustration that is mentioned by more than 70% of survey participants. Yes more than 70%. This seems hard to believe in this day and age of instant communication.

So how can you demonstrate that you do call back? Here are seven ways to do it.

1. Return all calls that go to voice mail within two hours on normal workdays.
2. Provide live phone coverage between 7:30 AM and 4:30 PM on normal workdays and if you don't have someone in the office – forward your calls to your mobile number.

3. Give your mobile number or your project manager's mobile number to clients and answer all calls between 7:00 AM to 5:00 PM on normal workdays.
4. Provide an email address for any-time communication and answer them once a day.
5. Set up a communications folder or message board at each jobsite that clients can use to provide information to the project manager. This is typically used by clients to leave notes for the next day and it avoids most overnight calls.
6. Ask your clients not to call after 5:00 PM except in a real emergency – this will eliminate most unnecessary after hours calls and allow you time with your family.
7. This is your communication policy so put this in writing and give it to each prospect during the first meeting.

Let me assure you that if you do these things consistently, your clients will rave about you and you will get more referrals and testimonials.

More importantly for this program – calling back prospects within a specified window will demonstrate that you are better than 70% of your competitors. Remember the survey – if 70% of the participants say it is a problem then 70% of the companies must be guilty.

This is an easy and very important advantage that you can gain over 70% of your competitors.

Another big frustration is not showing up on time – so make it a priority to show up on time. If you are running late call and let your prospect or client know this and never keep people waiting more than 15 minutes. This is a very big deal and one that can kill your chances before you even get started.

The remainder of the frustrations listed above can be easily overcome and we will discuss how to do that in later chapters.

Action Step – Create your company's written communication policy using the seven ways I shared with you. Add others or modify the ones I gave you but get it done today and start giving it to every prospect. These seven ways are the way YOU would want to be treated if the roles were reversed. By putting these in place you are thinking like a buyer.

A word of caution, if you do change your call back time – don't make it next day or 24 hours. That isn't going to make you stand out from your competitors.

Think like a buyer and it will be clear what your standards should be.

Be cautious about changing my recommendations because all of them are proven.

Chapter 4

Setting Yourself Up For Closing Success

If you want to get a warm reception from your prospect you need to set yourself up to succeed.

As I discussed in the previous chapter – communication is very important.

First impressions are lasting.

People seem to have forgotten that. In fact a trend I have noticed is that many contractors act like they are doing you a favor just by showing up.

So be polite, enthusiastic and positive on the phone. If you have staff that handle the initial contact, make sure you train them to be polite, enthusiastic and positive. Life is too short to deal with nasty or unhappy people.

Ask good questions to gauge how serious the prospect really is and to find out if this is a project that is likely to be profitable.

Too many times the business owner or sales person will run out to see every prospect that calls. I am sure you have had this experience - 2 minutes after you arrived at a prospect's location you realized that it wasn't a job you wanted.

Here are the six questions you should ask every time. You can ask others depending on your type of business or based on the answers you get. I will be focusing on marketing questions.

Naturally you should ask the technical questions that you will need answered. This is especially true for commercial and industrial opportunities.

Asking these six basic questions will help you to eliminate the tire kickers and price shoppers that are never going to hire you.

Spending five minutes on the phone will save you hours of wasted time on these prospects. You can devote the time you would have spent on them to do other things to build your business.

You may be thinking: "I don't have a full schedule right now so I want to go on every prospect call." I understand that but wasting two hours going on a call that you aren't going to get or is marginal at best is not the best use of your time.

I would suggest that you spend that time following up on other prospects, contacting people that have referred you, updating your website or a hundred other ways to generate more business.

How to Stand Out on the Initial Call with a Prospect

So let's go over the initial call and the questions you definitely want to ask.

First, keep your greeting simple.

"Thank you for calling XYZ Contracting (Services). This is Bill, how can I help you?"

Once they tell you – simply respond: "Great - I need to ask you some quick questions to gain a better understanding of your project – is that ok?" Now wait for their answer and when they say yes, proceed.

1. **"How were you referred to us?"** This will tell them that you get a lot of business by referral and it will also tell you what advertising is working because if they weren't referred to you, they will mention how they found you (yard sign, website, networking etc.).
2. **"Have you ever done business with us before?"** This is especially important if a staff person is handling the call as they won't know or remember every client.
3. **"How soon did you want your project completed?"** Don't automatically reject someone that needs the job done quicker than you can do it. Most reasons for wanting to get a job done are not that crucial. The caller typically wants to get it done quickly because *they finally decided* to do it NOT because it has to be done that quickly.
4. **"What is the primary reason that you want to get the project done?"** This is what is really motivating them to move forward and the answers you get here will help you to figure that out. What if they are selling the property and want a splash and dash job? What if they are having a wedding at the house

and time is crucial? In effect they are telling you "why" they want it done and that is far more important than what they want done.

5. **"Are we the first company that you have spoken to?"** Again, there are lots of reasons to ask this question. What if they say they already have three quotes in hand? What if they say they got such a strong referral that they called you first? What if they say they called three other companies and you are the first person that answered the call or called them back?

6. **"We like to hear from everyone who will participate in the decision so we get all the input we need – when would be a convenient time when everyone will be available?"** This will help you avoid this scenario: "I just have to go over this with my husband or my wife". When you hear this you will realize that they weren't the real decision maker even if they said they were.

Finally, close the call with a thank you, confirm the appointment and let them know that you will be sending them some important information for them to review before the meeting about your company.

How to Leave Your Competition in the Dust

I may be wrong but I don't know of even one company except our clients that sends information before the initial meeting.

All of our clients send information to prospects before they meet with a prospect.

This is a very important part of successful closing and will give you a significant advantage over all of your competitors.

There are two goals in sending information before the initial meeting:

1. Educate your prospect on what to expect in the process.
2. Provide the outline of all the good information you are going to share in the meeting.

So what should you say in the letter?

- First, keep it simple and to the point.

Some things to consider including:

- How long you have been in business
- If you are a member in a professional association for your industry
- List relevant certifications if appropriate

You will also want to show the details of the time and date of the meeting along with the person they are meeting.

It is also a good idea to describe what you plan to cover including answering their questions about your company and their project.

What else should you include?

This is an interesting question. Our clients have a specific package of information that they provide that we create for them that demonstrates trust, experience and respect for their property.

So think for a minute – what conveys trust? Proper licensing, adequate insurance, membership in professional associations...

What conveys experience? Testimonials, reference lists, project lists, certifications, awards...

In addition, many include company brochures and articles about the company.

Some also will include supplier's catalogs if appropriate.

Keep in mind that any competitor that uses that supplier or manufacturer's products will have the same catalog so sending that in advance is not something you typically want to do.

Providing this information will make your prospect believe that they can trust you. They will also be much more likely to ask for your recommendation on products and services.

Now you may wonder what percentage of people will actually review the information provided.

The answer we get from our clients is between 60% and 70%. So by sending information ahead of the meeting you get the jump on your competitors six or seven times out of ten. I will take those odds every time and you should too.

Another advantage of doing this is that when you get to your prospect meeting you won't have to spend a significant amount of time telling the prospect about your company and how great you are. You can cut right to the chase and start talking about THEIR project and their needs and isn't that what the prospect is really interested in?

The whole key is to set this up so you can easily get it out to each prospect. If you set this up ahead of time then it should only take two to three minutes to complete the letter with their address or to attach a PDF to an email if you are going to see them the next day.

Your competitors figure they don't need to send anything in advance – they run right out to see the prospect or they are confident they can sell everyone they meet in person. You may have done the same thing up until now. Now you know how to get the jump on your competitors.

Action Steps – Here are two action steps to put into place:

1 – Create your own prospect interview questions using the six questions as a base - you may want to add or slightly modify some of the questions but be careful if you do that because these are proven.

2 - Put together your letter and other materials that you will be sending to all your prospects before you

meet with them. Once you have it organized, send it to every prospect before the initial meeting.

Chapter 5

Getting Prospects Ready to Buy

Your prospects will already have a positive feeling about you by the professional way you answered the phone and from the information you sent in advance of the meeting.

That will put you two giant steps ahead of your competitors.

Now we are going to spend the next four chapters focused on the prospect meeting.

I am going to cover some groundbreaking tactics that our clients use day-in and day-out to win profitable business. The more of these that you do, the more successful you will become.

I am also going to cover some very important and sometimes obvious tactics. My apologies in advance since some of these will be common sense. Unfortunately, what we have found is that common sense isn't always common practice so bear with me. Also just because you know what you should do doesn't mean you do it every time.

I'll admit it I love to talk about my business and all our success.

Seriously, if you want to close more prospects you have to use your mouth in proportion to your ears for one-third of the time. I understand that for most people myself included this can be a challenge, especially if you are pumped up about your business.

What I have found is that you will close more prospects by cutting down the selling and increasing the buying.

What I mean by that is you need to focus on asking questions and letting them buy.

Most people are only going to retain 20% of what you tell them anyway so doing more of the talking doesn't help you as much as you may think.

You need to get good at asking questions and then listening to the answers.

You will be astonished at how many people will close themselves if you just ask them the right questions.

So Where Do You Start?

Remember the biggest frustrations prospects have from Chapter 3?

So show up on time and give yourself the time to comfortably complete the meeting based on the size of the project.

Rushing out to make another meeting isn't a great idea because it sends a message to the prospect that you have something more important to do.

Have a plan and keep the meeting focused and you will get good at finishing in the time you planned to spend in the meeting.

If you are going to be late – don't be more than 15 minutes late or you will be killing your chances. Think about it – how frustrated do you feel when someone is late? Your frustration goes up even more when they don't call to say they will be late.

This probably goes without saying but I will say it anyway. Be clean and neat. If you have a company shirt, coat or sweatshirt, make sure to wear it. It is also a good idea to keep a clean change of clothes with you in case you get dirty on a job.

I once had a painting contractor show up to my house in coveralls that had paint all over them and his truck was dirty and dented. He handed me a worn tri-fold about his company.

He had crossed out the address and written in the new one. I could go on about how poor an impression this guy made on me but I think you get the idea.

Naturally, none of these things gave me a lot of confidence and he had no chance to get the job.

The amazing part of this little story is that someone was hiring this guy.

I have met plenty of contracting and service companies that did some of these same things and yet they wondered why they were losing jobs to other companies.

You can beat out all of these people just by being neat, clean and professional. Let's move on to another important point.

Once you arrive – be alert – look at the condition of the property, the type of vehicles and their condition, how their property compares to others in the neighborhood or industrial park etc.

Look for opportunities to provide other services that the prospect may not have mentioned on the phone. Make a note to ask about these during the meeting.

Introduce yourself professionally and confidently. As soon as they introduce themselves, say: "I have some very important information to share and some follow-up questions to ask. Could we sit down for five minutes to do that?"

If they invited you in (this is likely) then you would simply say: "I have some very important information to share with you and I have some follow-up questions to ask, is that Ok?"

Now here are the opening questions

"Did you receive the package of information I sent?"

When they say, "yes", DON'T ask: "Did you have a chance to review it?" Most people will say yes even if they didn't because they will be embarrassed to admit they didn't review it.

So ask the question this way.

"What part did you like best? Or what part did you find to be most helpful?"

These are open-ended questions that will allow them to tell you what they like about you and your company! Isn't that cool? Now you can fill in the blanks and highlight what makes your company stand out from your competitors.

If they didn't review the information you sent. They will say: "Gee we didn't have time to review it." This will give you an opening to review the information you sent and you can highlight what makes your company stand out.

Action Step – Read this chapter again and practice asking the questions in your next prospect meeting. Also make a commitment to always be on time or to call if you are going to be late.

Chapter 6

Delivering the Number

Ok, if you cheated and you started here welcome to the program.

My recommendation is to go to the beginning because if you simply try to use this chapter you won't be anywhere as successful as if you follow all the steps.

When you follow everything I have covered up until this point you will clearly stand out from your competitors and getting to yes will be a whole lot easier.

Once you have gone through the early part of the meeting and reviewed the information about your company, you will need to gather the information you need to make your proposal. Be alert for opportunities to emphasize your strengths and to refer back to the information you already provided.

Next you will prepare your proposal and deliver it in on the spot or arrange to come back to deliver it in person.

In our experience – delivering the number in person will win you 20% to 50% more jobs. There are far too many reasons why this is true to cover in this program but several key reasons are you can answer additional questions, overcome objections and reinforce why you are the obvious choice.

There can be some circumstances when you just won't be able to get there to deliver the proposal in person – be prepared to have a lower success rate on those proposals.

If a prospect asks you to email the proposal, resist doing that because you are losing all of your leverage and letting them take control of the sales process.

Delivering the Number

Step 1 – Give the prospect their copy of the proposal and go over the project in whatever detail makes sense – *don't include the price on their copy* because they will immediately focus on that and not listen to anything you are saying – I can tell you from personal experience how frustrating it is for someone to go right to the price. By leaving it off you stay in control.

Step 2- Ask: "Now that I have reviewed our proposal and how we plan to successfully complete your project – do you have any additional questions about the project?"

Step 3 – If they say, "What is the price?", simply say: "If you are comfortable with everything that I have included in our proposal then we can review the price. If not we want to answer your questions first because any changes in the project will impact the proposal price."

Step 4 – Answer their questions.

Now you are ready to deliver the number – I am going to use an example for a painting contractor - naturally, you can fill in your industry:

- "Thank you for the opportunity to present our process and educate you about what to expect from a quality painting contractor – our complete price to paint your house *the right way* is $5,500 – are you ok with that?" Now you need to be quiet and let them speak first – they will tell you where you stand and now you are in a better position to overcome their objections or concerns.
- Let me emphasize the key words, "are you ok with that?" You are asking for a response and this is the key in negotiations. If they are ok – you got the deal – if they aren't this gives them an opportunity to voice their objections.
- If you just throw out the number then they may or may not respond. By asking if they are ok you are getting a go or at least you will know what objections you need to overcome
- Just using this approach will close more jobs for you.
- If they say, "Yes, I am ok with that." Then say: "Great, all I will need to hold a start date for your job is a deposit of 15%. Will you be using Visa, MasterCard or giving us a check?" (Naturally you will use the amount you require and the type of payment options you offer.)

What to Do If They Still Aren't Ready to Say Yes

If they voice some objections or say they are not comfortable then ask these questions.

After Delivering the Number If They Still Have Questions:

1. Ask them, "What questions can I answer for you?"
2. Respond to their questions or objections which will most likely be based on details, timing or price.

After You Have Answered Their Questions:

1. Ask them: "Does that answer all of your questions? Do you feel comfortable with our proposal? Does it seem reasonable?" People have a very hard time saying something isn't reasonable.
2. Depending on what they say – if they are ready to go – close the sale and get the agreement signed.
3. If they say they are comfortable: "I can get your spot on the schedule reserved today – Ok?" Now wait for their answer and close the deal.
4. If they say that you are a little high – you will need to be ready to overcome the price objection. I will cover that in more detail in Chapter 8.

I want you to realize that some people have a hard time making a decision. Even after you patiently answered all of their questions, there will be still some people that want to, "think about it". We will cover what you should do in this situation at the end of Chapter 8.

Once you start using what you are learning in this program, it will happen less and less.

Action Step – Read this chapter three more times and practice using the closing lines and questions. Get good at asking, "Are you ok with that?" Make a promise to commit all of these questions to memory. Don't worry if you forget one – your prospect will still be amazed at the professional way you handle yourself. They will never know that you bobbled a question.

Chapter 7

Four Words You'll Begin to Hear More Often: "When Can You Start?"

In chapter 5 I reviewed the opening of the meeting and the opening questions.

Once you have asked what the prospect liked about the materials you sent in advance of the meeting you will answer the same way every time – whether they answer yes and let you know what they liked or no they didn't review it.

This is your opportunity to quickly and efficiently review your materials with them. I am going to recommend that you put together the additional information I am now going to go over. Once you have it together - along with the information I recommended in Chapter 4 – you will hit the highlights during the meeting and then plan to leave it with them to review after the meeting or while you are taking measurements or writing up the proposal

This is one of the keys to, "Letting them buy".

What should you include? What should you cover?

Again, think back to Chapter 3 – you will want to put them at ease. Make them feel comfortable that you understand the biggest frustrations or fears they have and show how you overcome them.

In Chapter 3, I gave you some specific suggestions on how to overcome the biggest frustration, "They never call back".

Here is what I suggest you do right now – I am going to list the frustrations again. If I can summarize these frustrations – they focus on communication, respect and reliability.

Write them down and prepare a summary of how you overcome each of these frustrations. Be specific and then commit them to memory. I will give you some ideas on how you can demonstrate that you overcome them.

It would be even better if you put all of this in writing and give it to your prospect to review at their leisure. Simply say: "Here are our company procedures on communication, service and warranty. We have put this in writing so you can feel comfortable that this is what we do for all of our clients."

The Eight Biggest Frustrations Buyers Experience

In Chapter 3 you learned what the buyer's biggest frustrations are. For review:

1. They never call back
2. They never show up for appointments on time
3. They don't start or finish the job on time
4. It always costs more in the end then in the proposal
5. It won't look as good as they claim or work as well as they promise

6. They don't honor the warranty or guarantee
7. They don't finish the punch list
8. They desperately want to hire a trustworthy company but they don't know how to do that

Let's dive into these in detail and I will show you how you can easily overcome all of these frustrations.

Number 1 - *They never call back.* You overcome this by your actions, so call them back quickly. We suggest one or two hours tops. You should also give them a copy of your communication policy with the points I detailed in Chapter 3.

Number 2 - *They never show up for appointments on time.* If you showed up on time that proves it and if you didn't hopefully you called in advance!

Number 3 – *They don't start and finish the job on time.* Gather a group of testimonials that specifically address this issue. If you don't have those testimonials go ask every satisfied client to give you a testimonial on this. Even having three people say you start and finish on time is a lot better than none. Keep adding to this group of testimonials.

Number 4 – *It always costs more in the end than in the proposal.* I suggest you create a price guarantee for the work specified.

Notice how I said that. If the prospect wants to add to a project they should expect to pay for it. Be clear about the services you are providing for the quote given.

I realize that in certain circumstances you will need to include contingencies and this is especially true in remodeling if you are adding an addition for example.

Your prospect has a fear that they will get hosed in the end by a low up front price. Put it in writing and make it clear that you honor your price.

List the exceptions and keep them to a minimum if at all possible. This will be a big selling point against the low ball bidder.

Going back and asking for more money because you didn't do the best job possible in estimating is a sure way to kill referrals and possibly end up in a legal mess.

This is another frustration that can be backed up by specific testimonials that address this topic.

Here is a little story based on my personal experience.

> About ten years ago – before I started this company my wife and I decided to do an extensive renovation of the kitchen, family room and breakfast room in our house.
>
> One contractor that wasn't our first choice after the initial meeting gave us a list of 50 references (the other two gave us three references each). I decided to randomly call ten of his references and ask about him. Every one of the ten references I spoke with said they

would hire him again because he started and finished on time and most important charged them what he said he would.

This made the choice easy for us.

It was obvious that he was the guy. We hired him and it came out even better than expected. The fact that he gave me 50 references while the other contractors gave me three was a big buying point.

References and testimonials can make a big difference – get as many as you can.

Number 5 – *It won't look as good as promised or work as well.* Great pictures of all the different types of projects you have successfully completed are a must. It is a smart idea to hire a professional (maybe in a barter arrangement) to take the pictures.

Testimonials and references will also back this up. You can never have too many pictures. Make sure you have high end and lower end jobs. This will allow you to capture all types of projects.

Check out the local photography school and ask the teacher for the best students – some may be very happy to get paid to practice what they learned in class. This can be a very economical way to get great images.

I want you to realize that in remodeling and landscaping, your images are the best way to prove your expertise and creativity. In painting this is also important but a bit less so. If you are in a trade like plumbing and electrical, you can use images to prove your expertise by showing best practices or unusual projects.

Number 6 – *They don't honor the warranty or guarantee.* This can be a tricky one.

You don't want to go back to fix anything but stuff happens. Ironically, people that you make happy after they spot a problem are much more likely to sing your praises. When you fix the problem that is a great time to ask for a referral.

If you have been in business for 15 years – let your prospect know that you will be around to honor the warranty. If you are just starting out in your own business and have years of experience working for someone else – simply let them know that you spent ten years learning with a great company and that you have a solid business plan in place so you are confident that you will be around if something ever goes wrong.

You might also be able to say – in my ten years with XYZ we only had two callbacks on the 200 jobs I did or something similar as long as it is true.

Number 7 – *They don't finish the punch list.* This is similar to the last point and can be backed up using the same tactics. Here is a short list of some of the things you should provide to demonstrate that you are a trustworthy contractor:

- Registration with your governing licensing agency, if applicable
- Your current insurance certificate
- Membership in professional and civic organizations
- Certifications or evidence of specific training
- Reference list and testimonials
- Awards or published articles

Ok, I am going to assume that you put all of this information together in a folder. Let's review how to use this information – picture this - you have met the prospect and they have let you know whether they read the materials you sent before the meeting.

Number 8 - *They desperately want to hire a trustworthy company but they don't know how to do that.*

Here is one thing you don't want to do.

Don't go over all of this material at the beginning with the prospect. Highlight some of the stuff that makes your company great and then you can review other points during the rest of the meeting based on the answers you get to your questions or the concerns they voice.

The best way to do this is give them your folder – highlight a few key things and leave it with them to review at the end of the meeting or during the time you are preparing the proposal if you typically do that on the spot.

That is, "letting them buy".

Action Step – This is a very important chapter and you need to put together all of the information I suggested or put a plan in place to get it done. This will definitely make your company stand out from your competitors – even the very best.

Chapter 8

Successfully Overcoming the Price Objection and Other Common Objections

This is a very important chapter and one that you should plan to read many times.

Back in Chapter 1 – I gave the example that a golf professional can never practice every shot they will ever face because of the almost infinite places the ball can end up on a golf course. But they practice many similar shots so they don't panic when they face a challenging shot.

In the same way, you will occasionally be faced with a question or objection that you haven't heard before. Relax and simply think about a similar question or objection you have already answered.

This is called lateral thinking and it is the ability to adapt to a new but similar situation.

Almost all objections are expressed in terms of price and availability. The reality is that they may not trust you 100% or they can't figure out the real value of what you do or they don't believe you have enough experience to handle the project. So what you are really trying to communicate is trust and value.

In summary, the prospect wants to believe that when they hire you, they are getting someone that will do what they say they are going to do for the agreed upon price and in the end they will be satisfied.

They want to believe that by hiring you they have taken the risk out of the decision.

Now I will help you overcome the most common objections.

Overcoming the Availability Objection

One of the most common objections is availability. Here is how you can overcome this objection:

"I can appreciate that you want to get your project done ASAP because of the reasons you mentioned" (Refer to your notes at this point and list the reasons that they want to get it done ASAP). Then you can continue: "I have been in business for 15 years and I know the other quality companies in the area. We all talk about our businesses and the ones that I've spoken with are all booked three to six weeks out.

1. There could be a couple of reasons that the other company is available – they may be new in the business and need the business or they may do this as a sideline so they do it in their spare time. Unfortunately a number of people have entered our business due to the economy and they don't have the experience or training to complete your project properly.
2. We provide a start AND finish date. Unfortunately, some competitors will give you an attractive start date but they will only work

on your project for one day and then disappear for a week.

3. We don't want to make promises we can't keep. We would rather call you and let you know that we can do it earlier rather than later. There are some variables like the amount of rain we get or inspection delays but other than that we provide realistic dates that we will start and finish your project on. So aren't you willing to wait just a little longer to ensure you get a quality job?"

This is a very easy way to overcome this objection and this approach will win you more jobs.

Dealing with the Price Objection

Previously I recommended strongly that you present your proposal in person. The price objection is one of the reasons why this is so important. It is extremely difficult to overcome the price objection by phone or email.

The price objection is one that you will typically face even if the prospect is convinced that you are the best company for the job. It doesn't matter whether the economy is good or bad, most people will try to get a better deal.

Throughout this section I am going to be using a painting project in the examples. Simply changing out painting and inserting your industry will allow you to adapt this to your business.

Here is a common price objection: "We would really like to use you John but do you have any give in your price?"

Here is what you should use as a first response:

"We know every consumer wants the best deal – we also know that not every contractor adheres to a high standard, they only compete on price and in order to make a profit, they look for ways to shortcut the process because they don't expect to work for you again. We don't work that way – we get almost all of our work by referral and from repeat clients who ask us to maintain their homes and do additional work for them – isn't that the kind of contractor you want to work for you?" Now you want to wait for their answer.

If they are still stuck or don't look convinced you need to find out how big a difference they are talking about.

Let me give you an example – if this is a $20,000 job and they are looking for a $500 discount that is one thing but if they are looking for a $3,000 discount that is something completely different.

To find out how big a difference you are facing ask: "Do you have another proposal that you are seriously considering?" If they say yes then simply ask: "What is the difference in price that we are talking about?"

If it is a large price difference then you will need to educate the prospect that it isn't the same job. I advise my clients that any proposal that is 15% less than theirs means that the competitor has cut corners somewhere and you need to point out where they are doing it or are likely to do it to the prospect.

How to Overcome a Large Price Difference

If it is a large price difference here is what you need to say:

"Unfortunately some of our competitors will throw out a price that is unrealistically low just to win the job because they know that once the job is started it is prohibitively expensive to change contractors. They know that there will be plenty of opportunities to increase the price once the project is started.

In our business a 10% profit is considered excellent so when someone bids 25% less than us – they can't be doing the same job.

This is a large price difference and in our experience – they are not going to do the same job and here is why:

Our material cost is $2,210 and because we do a high volume of business with Benjamin Moore we get their best pricing and pass that along to you.

There are 150 hours of labor in your job to ensure your satisfaction and because we pay a competitive wage and provide health insurance for our guys that comes out to $11,250.

We are licensed and we pay for the state mandated insurance coverage. We work diligently to keep our overhead lean and that amounts to 20% of each project on average and that leaves us with about a 10% profit if everything goes right. Does that seem reasonable to you?"

If they say ok – then go right to the close and get them on the schedule. If they are still stuck and say something like: "Gosh John I understand but $3,000 is still a big difference."

Your best chance at this point is to review the other proposal. If the person trusts you they will let you review it. They want to get the best deal.

Say: "I would be pleased to review the other proposal with you. If it lines up with ours on materials, labor and warranty, I will wish you the best and we can part friends. If it doesn't line up I can give you the information you need to make the best decision. The worst that can happen is that you do end up getting a terrific deal. The best that can happen is that you will avoid starting a project that won't meet your expectations. If you are ok with that approach we can take the time to review the other proposal right now."

It should be relatively easy for you to point out where the other company has cut corners. If it looks like the same job as yours – be professional and thank them for their time. Let them know that if something does go wrong, they can call you.

This is a smart professional approach and our clients will attest that it has resulted in getting jobs when the other contractor screwed up.

One of my remodeling clients was faced with just this situation – their bid was $119,000 and the competitor's bid was $95,000. The client allowed him to review the other contractor's proposal and there was $15,000 of costs that didn't appear to be included in the total (permits, disposal etc.). He advised the prospect to ask the other contractor this question: "Is $95,000 the total price I will pay for this project?" After a bit of dancing the other contractor admitted that the total cost was more like $110,000. The client now knew that the other contractor wasn't being totally honest and went with my client despite the $9,000 difference.

Another Tactical Variation to Overcome the Price Objection

Here is one other tactic on the price objection that you can use especially on bigger differences:

"We have diligently worked to give you a price that allows us to meet your expectations and also meets the quality standards we have set. Our price is firm; we won't throw out a low number just to get a job. We take extra time to detail your quote so you won't be surprised at the end of your job. If you have established a maximum budget that you can spend and you want us to do the work – we can possibly adjust the size of your project to accommodate your budget but we won't compromise on quality just to get the job.

Unfortunately the reason some contractors can offer a low price is because they don't deliver much value - the most common complaints we've heard from potential clients center on hidden charges, companies that don't honor the warranty – even during the warranty period and companies that run off with client's deposits.

If you are comfortable with my company – I can get your spot on the schedule reserved today – Ok?"

Now wait for their answer and close the deal.

Closing the Sale for Small Price Differences

Now let's spend a little time on a small price difference - let's say $500 on a $20,000 job.

You need to decide how profitable the project is and if it makes sense to take the hit to get it. There are other reasons why it might makes sense to take it: the location, similar homes or buildings that need similar work on the same street, you have a gap in your schedule, the person can refer you to many people and *many more reasons that can support a reduced profit on this one job.*

If you decide to give in on the price difference then you need to get something back. That is just good negotiation. You can say: "Bill if you are comfortable with us as a company and we can set the start date now, I am willing to reduce my price by $500. Are you ok with that?"

You can also offer to meet them halfway. If any of the other reasons to take the job at the lower amount are in play – then going halfway is risky – if you lose a profitable job for $250 you will likely be kicking yourself all the way down the block.

If you really have cut this proposal to the bone and you can't budge then you will need to use the large price difference strategy or the next one that I am going to share with you.

This next strategy is a very, very powerful one and should only be used when everything else has been exhausted and they are still hung up on the price difference.

If the client says to you: "Gee the other contractor we like gave us a proposal that is $750 less than yours – do you have any give in your proposal?"

"Let me ask you something Mr. or Mrs. Smith, if we were the same price as the other company would you hire us?" Now wait for their answer – if they say no – pack up your stuff and go because it isn't about price – this is very unlikely.

When they say yes, ask: "What are the reasons you would hire us over them?" and they will tell you.

Your response will be: "That's why we are more expensive – it takes an investment of time and people to produce consistent quality systems and work – isn't that what you really want?"

When they say yes immediately say: "If you are comfortable – I can get you a spot on our schedule reserved today – Ok?"

Another price strategy that works especially with small differences:

Ask: "How long do you expect this paint job to look good?" Now you divide in your head the difference by the number of years they mentioned. Then say: "You may save $200 a year now, but if you have to repaint again in 3 years then it will end up costing you a lot more than the $600 you might save now. If you are comfortable with the value we deliver, I can get you a spot on our schedule today."

Ok let's cover the situation where you have been through the process, asked lots of great questions, answered everything the prospect can think of and they still want to, "think about it".

Say this: "Thank you for your time. I hope that you will consider ABC Painting to provide the level of quality service you expect and deserve. We look forward to working with you on this project.

May I call you in 3 days to follow-up?"

Wait for their answer.

When they say yes – ask for a specific time to contact them and confirm the number. "How about I call you at 6 PM on Thursday on your home number, does that work for you?" Wait for their answer. Finish with: "Great I will call you next Thursday at 6 PM – Here is my card with the time on it so you can put it into your calendar."

One last thought on this – don't make the mistake of asking: "Can I call you in a few days to follow-up?" That is too vague and then you will need to chase them.

Action Step - It takes practice to get good at overcoming objections. I encourage you to study this chapter again and again until you can deliver these scripts confidently in your own style.

Chapter 9

Know When To Follow-Up and When to Walk Away

It is always a good idea to send a thank you letter or note after meeting with the prospect for the first time and after any other significant meeting or part of the process.

Here are the goals you will want to achieve with a thank you letter.

You can reinforce the key reasons why you are the obvious choice to buy from and it also gives you one more chance to ask for the business.

It continues to build confidence in you and your company and sets you up as the most professional company the prospect has met. This is because no one sends a thank you letter so you will definitely stand out.

This is not an email.

Look at your mail for the next two weeks and count how many personal letters you get – this should convince you that sending a letter will definitely make your company stand out.

So what should you say in the letter?

Keep it simple and to the point.

Some things to consider including to reinforce why they should choose you:

- How long you have been in business and the number of similar projects you have completed
- If you are a member in a professional association for your industry
- List relevant certifications if appropriate
- Key things you do to ensure success
- Any key items that they mentioned that were important to them

If this sounds a lot like the letter you sent before meeting with them, it is. Think of this as bookends. The only real difference is that before you met with them you really didn't know that much about them or what they wanted. After meeting with them you should have a very good idea of what they want and what is important to them.

Now let's turn to a part of the process that most people dislike – following up with prospects that haven't responded or made a decision.

No matter how well you follow the steps in this program there will be times when you will need to follow-up with a prospect that wasn't ready to make the decision.

Several Scripts You Can Easily Adapt

Important Reminder – This is the same script you would be using face-to-face so if you are doing this on the phone remember to have your notes on overcoming the most common objections handy so you can confidently answer the price and availability questions – you shouldn't need to but why take the chance?

First I am going to cover how to handle the prospect if they have made a decision and later if they haven't made a decision. After that I will give you a script to leave on a prospect's voice mail.

I am going to use a painting contractor in this example and you should be able to see how easily you can adapt this for your business.

"Hi Mr. and Mrs. Smith this is John from ABC Painting

When we met on July 30th, you mentioned that you were looking to get started soon...I'm just following up to see if you've made a decision?"

If they say yes and don't tell you who they selected then simply ask, "May I ask who?" Now wait for their answer.

When they give you the name if you don't know the company say: "Wow...we've been in the painting business for 18 years and I am an active member of the Northern NJ chapter of the Painting and Decorating Contractors of America and I can't say that I've ever heard of them."

If you aren't a member of your professional association, then say: "We've been in the painting business in this area for 18 years and I am comfortable that I know all of the quality painting contractors and I can't say I have ever heard of them."

Then say: "May I ask, what were the primary reasons that you decided to go with them?"

Now wait for their answer.

"Thank you for your feedback." Then follow the Chapter 8 scripts for overcoming objections with one exception – if they are going with a much lower bid then go for broke and ask: "I appreciate that the other price is more attractive, I am still willing to review our proposals side by side so you can be assured that you are getting what you expect for that lower price. I can do that either tonight or tomorrow night if you are open to my offer."

If they are still convinced they made the right decision

"Let me leave you our number and feel free to call if they don't show up on time or something else goes wrong, we will be ready to complete your job."

What To Do If They Haven't Made A Decision Yet

Start with this: "Whatever decision you make, please consider what truly makes us stand out from our competitors.

We realize our industry is competitive and we appreciate this opportunity to earn your business. We've painted over 550 homes in the last 18 years so you can be assured that the project will be *done right the first time* without hidden charges, hassles or poor quality.

Did you have any questions about our proposal?" Now wait for their answer.

Use the script from overcoming objections to address their questions then ask:

"Does that answer your questions and are you comfortable with our proposal?"

If they are comfortable say: "I can get your spot on the schedule reserved today – Ok?" Wait for their answer

If they still aren't ready to give you a go - get a follow-up date.

"We hope that you will consider ABC Painting to provide the level of quality service you expect and deserve. We look forward to working with you on this project.

May I call you on Friday to follow-up?" When they say yes – make a specific time.

The key here is to set a firm date and time for a follow-up. Avoid saying something like: "Can I call you in a couple of days?"

The bigger the job the more likely that your prospect will avoid making the final commitment until that last minute – this is just human nature so be prepared.

Now I am going to cover the situation that you get voice mail when you call to follow up:

"Hi Bill this is John from ABC Painting and my phone number is (855) 666-1122.

When we met on July 30th, you mentioned that you were looking to get started soon…I'm just following up to see if you've made a decision.

We realize our industry is competitive and we appreciate this opportunity to earn your business. We've painted over 550 homes in the last 18 years so you can be assured that the project will be *done right the first time* without hidden charges, hassles or poor quality.

Please call us at (855) 666-1122 with any questions you have. If we don't hear from you by Friday we will give you one additional follow-up call."

That wraps up the section on follow up. If you do everything right in the process, these calls should be a lot easier since you should have covered all of the objections in your meetings with the prospect.

When to Walk Away

Now let's turn to another situation that keeps many people up at night – when to walk away.

It can be very tempting to take on every prospect that says yes – especially if you are slow; but if you have a bad vibe about a prospective client it is ok to walk away. Some people will make the wrong decision and not hire you. Maybe they don't have the money or maybe their brother-in-law is in the business and they were just trying to keep them straight.

If you have followed the advice in this program and given your prospect all the education and information to make the best decision and they still don't get it – let them go – they will very likely be the kind of client you don't want anyway.

There are many reasons that will serve as warning signs about a prospect.

Let's go over the most important ones:

- The prospect is overly focused on price. This is a prospect that will nickel and dime you and look for free services. This is also an indication that they haven't paid attention to all the good information you shared about your company so they think they can hire anyone and get the same result. Avoiding this prospect is a good idea.
- The prospects can't seem to agree on anything. This is especially true with spouses. If they aren't on the same page on the project then you are going to make one of them unhappy and that is another recipe for disaster.
- They keep changing their mind about the project. This is a sign that they will do the same thing in the middle and changing a project in the middle can be very expensive and they won't want to pay for it.

- The prospect is just plain nasty or negative. This prospect will not be happy with anything you do. Life is too short to deal with people like that. Save yourself the pain and bow out gracefully. Tell them you have too much on the table to deliver what they want or anything else that will allow you both to save face.

A final thought on this. We have all taken on projects that we had some doubt about and then regretted it after we were committed. If you do get stuck, finish the project as promised and keep your standards high. In the end you will be proud of yourself.

Action Steps – Here are three action steps to put into place:

#1 - Put your thank you letter together and send it out after the initial meeting and after any other significant meeting.

#2 - Set up your follow-up script and keep a copy with you at all times. This will allow you to make your calls when you are out of the office. Set up your follow-up schedule so it becomes part of your routine every day. Too many contractors send out their proposal and hope for the best. Make your follow-up calls and you will stand out from your competitors and win a lot more jobs.

#3 - Learn how to say no gracefully. No is one of the most empowering words you can use. Avoid jobs that will frustrate you because that will have a negative impact on your whole business.

Summary Part 2

Your Sales Closing Success Implementation Plan©

You now have an implementation plan that will propel your company to levels of success you have only dreamed about.

You know specifically how to make your company stand out and it will be obvious to your prospects. It will be much easier for you to think like your buyer.

All you have to do is put it in place. The action steps are designed to help you to make consistent progress.

I can assure you that once you have all of these steps in place you will be in total control of your sales process and your business.

When you do, your competitors will be wondering what you are doing to be so successful.

Don't tell them but be sure to tell your friends about my book and program.

It has been my pleasure to give you this blueprint for success.

Bonus Chapter

Marketing's New Rules™ of Closing Success

I spend most of my days listening and observing. What I have found is that many common sense principles are just as relevant today as they were 20 years ago. Other principles are still relevant but technology has forever changed how they are applied. This is the essence of Marketing's New Rules™.

It's easy to get caught up or sidetracked by the latest trends and technology - blogs, Twitter, Facebook, LinkedIn, iPad, iPhone, Droid devices…the list is endless.

Marketing's New Rules™ is not about the newest technology, tricks or gimmicks.

It's all about communication and education.

This is still the number one issue with contractors and service providers and it doesn't need to be because this is easy to fix.

Rule #1– Answer Your Phone and Return All Calls

I spent considerable time in previous chapters covering this frustration and ways to overcome it.

If this rule sounds ridiculously simple consider these facts:

- 70% of participants in a recent survey of buyers that have hired a contractor said: "They don't return my calls promptly".
- In another survey, only 56% of home remodeling or home renovation companies had a live person answer the phone.
- Participants in another survey were asked what does the phrase "phone calls returned promptly" mean to them and the answers ranged from five minutes to twenty-four hours and everything in between.

So here is a key idea - Just answering your phone and returning all calls will generate more business because 70% of your competitors aren't doing that.

How is it possible in this age of instant communication that companies can't figure out how to answer their phone and return calls promptly?

So how can you distinguish yourself from your competition?

In Chapter 3 I gave you seven ways to overcome this issue. Here are four more ways that you can use technology to help you:

1. Use an answering service. They can quickly screen your calls into prospects and clients. They can then send you an email or text message with the subject line – "prospect" or "client".
2. Forward your office phone to your mobile number.

3. Set up a separate phone line that you only use in ads so you know when that phone rings, it is a prospect calling.
4. Finally, set a goal to return all calls within a specific time period: one hour, two hours, morning by noon, afternoon by 6 PM etc. and then rate yourself on how well you implement this each week.

Let me repeat this.

Your customers and prospects will be amazed at your responsiveness and just answering your phone and returning all calls will generate more business because 70% of your competitors aren't doing that.

Action Step – I recommend that putting your communication policy in place should be your first priority. If this is the only strategy or tactic that you put in place from this book it will win you more business.

Rule #2 – How to Attract The Low Price Shopper – Guaranteed!

If you want to attract the low price shopper do these two steps:

1. Meet with the prospect for as little time as possible or not at all
2. Simply complete your proposal with a price and mail or email it

At best, the only people that will hire you are the low price shoppers and at worst, this is a complete waste of time because you are far less likely to close the sale.

Now I know many of my readers are probably thinking: "Hey we know that and we don't do that."

Ok but here is what we found working with some very successful clients:

- They didn't have a sales process in place that they followed consistently
- The busier they got, the more likely that they would only do the two steps above and hope for the best

Ok I don't always follow my diet and exercise plan either.

So what can you do to fix this?

Set up a sales process that works for your size company.

For example: create a lead or prospect sheet and include the key questions to figure out if this is a lead you want to pursue (hint: you can make money on it.)

If you get a call from someone 30 miles away that wants you to: paint two rooms or cut their lawn or patch a small roof leak or some other small project – my advice is don't waste your time unless this is a referral.

Just screening out the deadwood up front will save you hundreds of hours a year.

You can use that time to work smarter at getting better leads and closing them and that means more time to deliver that proposal in person. There are so many advantages to delivering the proposal in person yet most contractors drop it in the mail and hope for the best.

Have a system and you can easily knock them out of the box

Just being busy doesn't translate into success. People sometimes think that since they are busy they are effective and making money. Maybe they are but I have met a lot of busy people that weren't making much money and they didn't have time for themselves either.

Following a system will make you more efficient and help you close more business because your prospects will realize you are organized and many of your competitors aren't.

Let me repeat this...

Your customers and prospects will be amazed that you took the time to deliver your proposal and answer their questions. It's also a lot tougher to tell someone "no" in person.

Action Step – Put your communication plan in place and practice screening prospect calls. If you do decide to go on a marginal call – use all the material in this program and make it a personal challenge to close the deal for a nice profit.

Rule #3 – The Buyer is Now In Control of the Sales Process

- Today most marketing is "inbound" - meaning buyers search for information about products or services as soon as they determine that they have a need and long before they contact anyone.
- Buyers can easily find alternative providers of your services including someone to replace you using Google and other search engines and they can do it in seconds.
- Once buyers find you through whatever lead generation strategies or tactics you are using, success in closing is based on how much quality information prospects can find out about your company. What they find determines if they will contact you.
- If a prospect is unable to find quality information about your company they will eliminate you from the selection process and won't even consider contacting you.
- Quality information includes images.

I will now review examples of how to make your online content a lot stronger.

Any place on the web that a prospect can find information about you should be managed to your advantage. Some places like your website are easier to control. Some rating sites can pose significant challenges if negative content is posted.

The best place to start is your website. Make it an asset instead of a liability.

Think about all the great evidence of success that you gathered, created and updated in Chapter 4. A prospect should easily find all of this on your website and you need to help the prospect understand the value of each point you make.

Here is an example about awards. Many people that have received a Contractor of the Year award from NARI simply mention or list it. There is so much more you can do.

I guarantee you that most people have no idea the effort that goes into submitting a project for this type of award so put it in perspective.

Every remodeling contractor in your area could compete for this type of award so when you talk about it put it in perspective. "Out of approximately 750 remodeling contractors in Union County, ABC Remodeling is the only one to be recognized for this award. The award is given to remodeling contractors that demonstrate excellence in functionality, problem solving, aesthetics, craftsmanship and innovation in the construction industry."

Use the total number of contractors in your area because they could all compete for it if they joined NARI. It is their choice to join or not to join and it is also their choice to compete for the award if they are a member.

I find it interesting that Angie's List is finally saying what percentage of contractors are awarded the Super Service Award. We have had our hand slapped for doing just that for our clients in the past.

Action Steps – Here are two action steps to put into place:

#1 – Update the quality or your online content. Make it part of your monthly plan to continue to update and improve your online content.

#2 – Look for opportunities to win awards and recognition for excellence in your industry. When you post your awards put them in perspective for your prospects.

Rule #4 – Images and Video Tell Your Story and are Skyrocketing in Importance

Most contractors and service providers haven't figured this out.

You can gain a significant advantage over you competitors by posting quality images and videos.

It doesn't matter if you are in remodeling or plumbing, you can use quality images and video to crush your competition and they probably won't even know why.

For some businesses the need for quality images is obvious – like home remodeling, construction, landscaping, marine construction or any industry where there is a big difference before and after the project.

For other businesses like plumbing or electrical the need for quality images or video might not be so obvious. For those industries focus on best practices and "how to" and "how it's done" videos to demonstrate your expertise.

This is a large topic that could fill an entire book so I will simply hit the highlights.

I have reviewed hundreds of websites that had dark, blurry or out of focus images. Other sites have images that don't show the details that make the project stand out.

All of these issues send a clear message to the prospect about the quality of your work AND it isn't the right message.

This is supposed to be the representation of your best work and your expertise.

Action Step – Use quality images and video to bury your competition.

Here are some basic things to keep in mind:

- The more important the project, the more important it is to have a professional take your images – good quality equipment won't

guarantee high quality images in the hands of an amateur
- Only post quality images
- Don't use stock photos
- Take before and after shots from the same angle
- The more important the project, the more photos of that project you should include – five to ten is usually a good number but it could easily be more for say a whole house renovation – this is how you can tell the story of your project and the way you work
- Keep adding new images to your portfolio
- Post images that depict the right way to do things
- Tag each image for SEO purpose – for example – Interior Painting – Union or Kitchen Remodel – Santa Clara
- Make use of YouTube because it is on its way to becoming the most searched website
- Create short "how to" and "how it's done" videos to demonstrate your knowledge and expertise
- Create case study videos that tell the whole story about important projects
- Tag your videos properly and then embed them on your website - this is a bit technical and there is plenty of free instruction on the web or you can ask just about anyone in college to help you.
- If you have a showroom – create a video to tell your story
- If you have a workshop – create a video to demonstrate your expertise

- People love taking and seeing images and video so give them something to go "wow"
- Pay attention to changing trends so you can take advantage of them in your business

As I said at the beginning, most contractors are still doing things like it was 1985 so the market is wide open for you.

This is Mike Jeffries and I wish you great success in your business and in your life.

The Next Steps

Remember:

The Fastest and Easiest Way to Grow Your Business Is To Start Closing More of the Leads You're Already Generating!

I urge you to keep working and learning how to improve your Closing Success.

I offer a number of options and programs to get you started.

- Sign Up for my FREE weekly Closing Success Quick Tip©. Every Monday you'll get a 90-second tip that will help you close more business and generate more leads.

- Subscribe to my FREE Podcast. This is not just about me. It's all about you and growing your business. I interview some of the most successful contractors and pick their brains about what they're doing to "crush it" plus I interview industry experts who share best practices.

- I have a nice resource of FREE downloads of other material on our website: Closing Success System Downloads.

And if you're ready to get serious about mastering your ability to close just about any job you want at the price you deserve, you'll want to take a look at these two programs - The Contractor's Closing Success Formula© and the Closing Success System© and choose the program that best fits your situation.

Hundreds of contractors throughout North America, Australia, South Africa and even Poland have taken what they learned in one of these two programs to increase their closing success by 50 to 200%!

The Contractor's Closing Success Formula©

This Online Training Program Will Teach You Step-by-Step What You MUST Do to Double, Maybe Even Triple, Your Closing Success Rate

You now have the knowledge and ability to confidently close more sales as long as you study the material I have provided and put it into place.

We understand that you're trying to run your business, manage crews, inventory, payroll, cash flow, materials...

What I've found is that some people want more help. They love the material and they are self-starters.

I have a great solution that we launched in 2011 - The Contractor's Closing Success Formula©.

This online training program lets you work on building your own Closing Success System at your own pace.

If you invest a little time in yourself and your business you'll find that you'll be closing more jobs and generating more leads.

In the Contractor's Closing Success Formula©, I've laid out specific, step-by-step training modules for you to follow. Some highly motivated clients have implemented all the tools in two weeks. Most put in six to eight hours a week and complete all the tools in four to six weeks.

You have unlimited access to the program for as long as you need. *There is no time limit.*

I want you to realize that you won't have to wait to complete all the tools or have everything perfect to start using the system.

In fact once you complete Module 2 you will have one of your most powerful closing tools ready to go.

Using audio, video, worksheets, training guides and ready-to-use template documents, you'll be creating the same Closing Success Tools© that I create for my one-on-one clients.

Everything that I include for my one-on-one consulting clients is in here and nothing has been left out. You will be able to ask me questions and get the right answers just in case you aren't sure what to do.

This is a cost effective way to benefit from the collective success my clients have enjoyed over the years. It's all in here.

I won't mislead you – This is NOT an overnight quick fix. Although many, many, of our clients closed the very first prospect they used the system with (even if they didn't have every tool perfect).

Two-Week FREE Look Period

Once you have signed up for this program, you'll have 14 days to review the foundation materials and jump into the first two modules of the program.

If you don't think it's right for you or your business, I will gladly refund the cost of the program.

My goal is to help you reach the success you deserve. I am not in business just to sell products and services.

If you're a good, reliable, ethical company that's dedicated to providing quality work at a fair price...if you can read, listen and follow directions...if you're committed to rolling up your sleeves and putting a little bit of time into this program in the coming weeks...then I'm confident that you can experience the same kind of results my consulting clients enjoy.

Personal, Customized, One-On-One, I Do It All For You Consulting

The Closing Success System© Will Give You the Tools and Training So You Can Close More, Charge More and Make More Money Now

If you don't think you have the time or the discipline to complete our online program, then consulting would be the way to go.

The Closing Success System© has helped our clients generate tens of millions of dollars in increased sales every year for the past six years. Our one-on-one training will put you over the top.

You can start making a lot more money and it takes as little as 30 days to *fully* implement and we frequently get it done sooner than that. Many of our clients see results in the first two weeks simply because they go for it...and you can too.

One other thing - we are with you every step of the way. WE customize every piece to your business. WE train you how to quickly get it in place. WE do all the heavy lifting and you reap the rewards with a minimum amount of time on your part.

The materials are all done and ready to go – the system is totally scalable whether you have two guys or two hundred guys and whether you have no one in the office or four people in the office. We realize that most companies don't have someone in the office so we designed the system to be virtually automatic.

If you'd like to explore this option, please contact me directly and I'll set up an initial business evaluation call to discuss whether or not this would be a good fit for your business and your goals.

Mike Jeffries
(877) 280-0715
Mike@ClosingSuccessSystem.com

Don't expect any kind of sales pitch – we work the same system we teach you to implement.

So decide which program will work best for your business.

- Choose our online program the Contractor's Closing Success Formula© and put the system together for yourself.
- Let us create all of your tools and train you how to easily implement the Closing Success System©.

It's your choice and there is one thing we are confident of regardless of which program you choose. Once the tools are implemented, you will be on your way to closing more profitable sales for your company.

Resources

I've listed some of the businesses and organizations that I've worked with over the years that deliver real value to the construction industry. *These are the Good Guys!* If you know of a company that should be on this list, please let me know!

Markup and Profit
http://www.markupandprofit.com

Michael Stone, author of the book, *Markup & Profit, A Contractor's Guide Revisited*, has helped thousands of contractors build stronger, more profitable businesses.

His company, Construction Programs & Results, Inc., provides products and services designed specifically for construction-related businesses of all sizes, whether general or specialty contractors involved in new home building, remodeling or commercial construction.

Michael Stone is my recommended coach for our contracting clients. His knowledge and practical experience dwarf other experts in his field.

NetLZ Consulting
http://www.netlz.com

Moshe Zchut and his wife and business partner Sigalit Amir-Zchut and their entire team approaches each project with creativity, innovation and a "get it done now mentality".

NetLZ specializes in web marketing including: web design, search engine optimization (SEO), search engine marketing (SEM), web advertisement, web 2.0, social media optimization and other digital marketing techniques.

Having a website is just a start. NetLZ is your web marketing company. They will help you bring more traffic to your site and that translates into more sales.

NetLZ is my recommended web design and web marketing company. I have successfully collaborated on websites for contractors in: remodeling, commercial construction, landscaping, painting, roofing, windows, plumbing, IT consulting, property management and more in the commercial and residential markets.

MyOnLineToolbox
http://www.myonlinetoolbox.com

MyOnlineToolbox helps contractors get more work and manage the job! A service is available to learn how to get more Free Leads to your Contractor Website. And internet software is available to help create and track estimates, invoices and money.

The contractor software is designed by successful contractors who want a simple, easy to start and inexpensive solution to run their business. The education is designed for the owner to have a better understanding of their own online marketing

Map Business Growth Strategies
http://www.MapBusinessGrowth.com

Patrick Walsh has been my, "go to guy" for almost ten years when it comes to sales and marketing copy for my clients *and* myself.

He's one of the few pure copywriters I've run across that has years of experience working with businesses of all sizes in the construction, home remodeling and home services industries. He knows the industry and he knows your prospects.

www.ingramcontent.com/pod-product-compliance
Lightning Source LLC
Chambersburg PA
CBHW071720170526
45165CB00005B/2084